NATIONAL GEOGRAPHIC KIDS

美国国家地理

Sharks

鲨鱼

懿海文化 编著

马鸣 译

第三级

外语教学与研究出版社
FOREIGN LANGUAGE TEACHING AND RESEARCH PRESS
北京 BEIJING

京权图字：01-2021-5130

Original title: Sharks
Copyright © 2008 National Geographic Partners, LLC. All Rights Reserved.
Copyright © 2022 Bilingual Simplified Chinese/English edition National Geographic Partners, LLC.
All Rights Reserved.
NATIONAL GEOGRAPHIC and Yellow Border Design are trademarks of the National Geographic
Society, used under license.

图书在版编目 (CIP) 数据

鲨鱼：英文、汉文／懿海文化编著；马鸣译 . —— 北京：外语教学与研究出版社，
2021.11（2023.8 重印）
（美国国家地理双语阅读 . 第三级）
书名原文：Sharks
ISBN 978-7-5213-3147-9

I. ①鲨… Ⅱ. ①懿… ②马… Ⅲ. ①鲨鱼－少儿读物－英、汉 Ⅳ. ①Q959.41-49

中国版本图书馆 CIP 数据核字 (2021) 第 236732 号

出 版 人　王　芳
策划编辑　许海峰　刘秀玲　姚　璐
责任编辑　姚　璐
责任校对　华　蕾
装帧设计　许　岚
出版发行　外语教学与研究出版社
社　　址　北京市西三环北路 19 号（100089）
网　　址　https://www.fltrp.com
印　　刷　天津海顺印业包装有限公司
开　　本　650×980　1/16
印　　张　37.5
版　　次　2022 年 3 月第 1 版 2023 年 8 月第 4 次印刷
书　　号　ISBN 978-7-5213-3147-9
定　　价　188.00 元（全 15 册）

如有图书采购需求，图书内容或印刷装订等问题，侵权、盗版书籍等线索，请拨打以下电话或关注官方服务号：
客服电话：400 898 7008
官方服务号：微信搜索并关注公众号"外研社官方服务号"
外研社购书网址：https://fltrp.tmall.com

物料号：331470001

Table of Contents

CHOMP!

What is quick?
What is quiet?
What has five rows of teeth?
What glides through the water?
CHOMP!
It's a shark!

Sharks live in all of Earth's oceans.
They have been here for a long time.
Sharks were here before dinosaurs.

OCEANIC WHITETIP SHARK

CARTILAGE: Cartilage is light, strong, and rubbery. Shark skeletons are made of cartilage.

Shark tail fins are larger on top. This helps them move through the water better.

HAMMERHEAD SHARK

A shark is a fish. But a shark is not like other fish. Sharks do not have bones. They have soft cartilage instead. Cartilage helps sharks twist and turn. Cartilage helps sharks move and bend.

If a shark loses a tooth, a new one moves forward to take its place.

Shark skin feels bumpy and rough. It's hard like sandpaper. It protects sharks and helps them swim faster.

Shark Pups

Shark babies are called pups. Some pups grow inside their mothers. Other pups hatch from eggs.

LEMON SHARK

Lemon shark pups grow inside their mothers. The lemon shark mother goes to shallow water to give birth. The pups stay near the shallow water until they are grown.

These fish are called remoras. They hang around sharks and eat their leftovers.

LEMON SHARK PUP

MERMAID'S PURSE

Swell shark pups hatch from eggs. The mother sharks lay the eggs in hard cases. People call the case a mermaid's purse.

A pup coming out of its egg case

SWELL SHARK PUP

Swell shark mothers lay up to five egg cases at a time. In nine months, the swell shark pups are born.

Pups Grow Up

NURSE SHARK

PREDATOR: An animal that kills and eats other animals

PREY: An animal that is killed and eaten by another animal

When shark pups grow up, they are awesome predators. They have many ways to sense their prey. Did you know a shark can smell blood from miles away? It can smell one drop of blood in 25 million drops of ocean.

Sharks can see better than humans can. Even in deep, dark water, a shark can see its prey.

Sharks take a test bite of prey before eating. Their taste buds tell them if the prey is fat enough to eat.

What is a great white's favorite candy?

 A jaw-breaker.

GREAT WHITE SHARK

15

What Big TEETH You Have

SAND TIGER SHARK

Sharks have many rows of teeth.
They are always losing some teeth.
They are always growing new teeth.
A shark uses up more than 10,000
teeth in its life.

Different sharks have different teeth.
Their teeth are perfect for what they eat.

Long, spiky teeth are for catching.

Flat teeth are for grinding.

Serrated teeth are for ripping.

WORD BITE

SERRATED: When something is serrated, it has a jagged edge, like a saw blade.

WORD BITES

PREHISTORIC: Prehistoric is a time before people wrote things down.

EXTINCT: A type of plant or animal no longer living

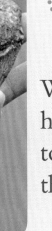

MEGALODON TOOTH

Wow! Prehistoric sharks had really big teeth—up to about six inches! Good thing these guys are extinct.

The megalodon is a prehistoric shark. Scientists made a life-sized model of the megalodon's jaw and put in the teeth they have found. You can imagine how big the shark must have been.

Imagine This!

A giant shark is gliding through
the water.
A swimmer is nearby.
The shark gets closer.
It is huge.
It opens its giant mouth and…

WHALE SHARK

…sucks in a big mouthful of water.
The swimmer is fine.
The shark is a whale shark.
Whale sharks are the biggest sharks.
But they have tiny teeth.
They eat tiny animals called plankton.

Blue-Ribbon Sharks

WEIRDEST
The Hammerhead Shark

A hammerhead shark has a head shaped like a giant hammer. Its wide head is great for hunting.

The dwarf lantern shark is about eight inches long. It has a glow-in-the-dark belly.

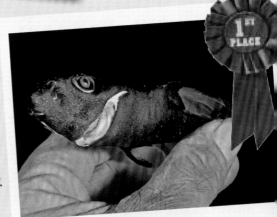

SMALLEST
The Dwarf Lantern Shark

CREEPIEST
The Great White Shark

When a great white bites its prey, its eyes roll back into its head. This protects its eyes.

FASTEST
The Mako Shark

The mako is the fastest shark. It can swim up to 20 miles per hour. Makos leap clear out of the water to catch prey.

23

Now You See Them...

DWARF LANTERN SHARK

Some sharks glow in the dark! Do you see something shiny in the water? Watch out! The tiny dwarf lantern shark is covered with a glow-in-the-dark slime.

The dwarf lantern shark is a deep-sea shark. Many deep-sea animals glow. Scientists think glowing might help predators attract prey.

Most sharks are hard to see. They have a dark back. From above, they blend in with the water. They have a white belly. From below, they blend in with the sky.

WOBBEGONG SHARK

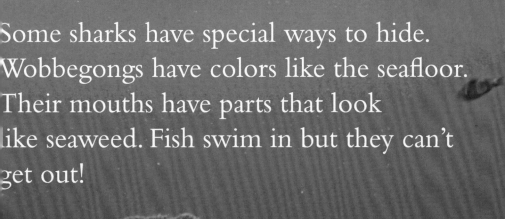

Some sharks have special ways to hide.
Wobbegongs have colors like the seafloor.
Their mouths have parts that look
like seaweed. Fish swim in but they can't
get out!

Shark Attack!

One day Bethany Hamilton went surfing. Suddenly, a tiger shark attacked. It tugged her as she held onto her surfboard. It took a big bite out of her surfboard. It also took Bethany's left arm.

After the attack, Bethany wanted to keep surfing. She is not afraid to go in the water. She knows that shark attacks are rare.

Bethany says, "One thing hasn't changed—and that's how I feel when I'm riding a wave."

People Attack?

Shark attacks are scary, and terrible. Sharks can be a danger to people. But people are a bigger danger to sharks. Millions of sharks die in nets set to catch other fish. Others are killed on purpose.

Many types of sharks may become extinct. Sharks have been on Earth for billions of years. Sharks and people need to learn to share the sea.

GRAY REEF SHARK

Glossary

CARTILAGE: Cartilage is light, strong, and rubbery. Shark skeletons are made of cartilage.

EXTINCT: A type of plant or animal no longer living

PREDATOR: An animal that kills and eats other animals

PREY: An animal that is killed and eaten by another animal

PREHISTORIC: Prehistoric is a time before people wrote things down.

SERRATED: When something is serrated, it has a jagged edge, like a saw blade.

参考译文

▶ 第 4—5 页

大嚼特嚼！

什么动物速度快？
什么动物很安静？
什么动物有五排牙齿？
什么动物在水中滑行？
大嚼特嚼！
是鲨鱼！
鲨鱼生活在地球上的各大洋里。它们已经存在很久了。鲨鱼早在恐龙之前就已经存在了。

远洋白鳍鲨

▶ 第 6—7 页

小词典

软骨：软骨很轻，很坚韧，富有弹力。鲨鱼的骨骼是由软骨构成的。

鲨鱼是鱼类。但鲨鱼和其他鱼不同。鲨鱼没有骨头。不过它们有柔软的软骨。软骨让鲨鱼能急转弯。软骨让鲨鱼能曲身活动。

鲨鱼尾鳍的顶部很大。这有利于它们在水中灵活移动。

双髻鲨

如果鲨鱼掉了一颗牙齿，新的牙齿会前移，补上它的位置。

鲨鱼的皮肤表面凹凸不平，十分粗糙。它像砂纸一样硬。它保护鲨鱼，让它们游得更快。

33

▶ 第 8—9 页

鲨鱼幼崽

鲨鱼宝宝叫"幼崽"。有些幼崽在妈妈的肚子里长大。另一些幼崽是从卵里孵出来的。

柠檬鲨幼崽在妈妈的肚子里长大。柠檬鲨妈妈到浅水区生产。柠檬鲨幼崽在长大之前一直待在浅水区附近。

这些鱼叫"鲫鱼"。它们跟着鲨鱼，吃鲨鱼的"剩饭"。

柠檬鲨

柠檬鲨幼崽

▶ 第 10—11 页

"美人鱼的钱包"

从卵鞘里游出来的幼崽

绒毛鲨幼崽

绒毛鲨幼崽是从卵里孵出来的。鲨鱼妈妈将卵产在坚硬的鞘里。人们把这种鞘叫作"美人鱼的钱包"。

绒毛鲨妈妈一次最多产五个卵鞘。九个月后，绒毛鲨幼崽出生了。

幼崽长大了

小词典

捕食者：杀死并吃掉其他动物的动物

猎物：被另一只动物杀死并吃掉的动物

护士鲨

当鲨鱼幼崽长大后，它们就成了可怕的捕食者。它们有很多种方法觉察到猎物。你知道吗？鲨鱼能闻到数英里（1 英里约等于 1.61 千米）以外的血。它能在 2,500 万滴海水中嗅到 1 滴血。

鲨鱼的视力比人类的好。即使在黑暗的深水区，鲨鱼也能看到猎物。

在吃之前，鲨鱼要先尝一口。它们的味蕾告诉它们猎物吃起来是否足够肥美。

大白鲨

你的牙齿好大啊

沙虎鲨

鲨鱼有很多排牙齿。它们总是在掉牙。它们又总是长出新牙。一条鲨鱼一生会有超过 10,000 颗牙齿。

▶ 第 18—19 页

不同的鲨鱼长着不同的牙齿。它们的牙齿非常适用于它们的食物。

又长又尖的牙齿用来捕猎。

扁平的牙齿用来咀嚼。

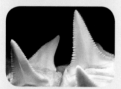

锯齿状的牙齿用来撕扯食物。

小词典

史前的：史前的是指在人类把事情记录下来之前的时间段。

灭绝：一种植物或动物不再存在

巨齿鲨的牙齿

哇！史前鲨鱼的牙齿非常大——可达约6英寸（约15.24厘米）！幸好这些家伙已经灭绝了。

巨齿鲨是一种史前鲨鱼。科学家做了一个巨齿鲨颌骨的等身模型，把他们发现的牙齿装在上面。你可以想象一下这种鲨鱼有多大。

小词典

锯齿状的：如果某个东西是锯齿状的，那么它的边缘是凸凹不平的，就像锯条一样。

▶ 第 20—21 页

想象一下！

一条巨大的鲨鱼在水中滑行。
一个游泳者在附近。
鲨鱼越来越近。
它很大。
它张开大嘴，然后……
吸了一大口水。
游泳者好好的。
这种鲨鱼就是鲸鲨。
鲸鲨是最大的鲨鱼。
但是它们的牙齿很小。
它们吃一种名叫"浮游生物"的微小动物。

鲸鲨

鲨鱼之最

第1名

最奇怪的
双髻鲨

双髻鲨的脑袋就像一个巨大的锤头。它那宽宽的脑袋非常适合捕猎。

第1名

最小的
矮灯笼鲨

矮灯笼鲨约8英寸（约20.32厘米）长。它的腹部可以在黑暗处发光。

第1名

最恐怖的
大白鲨

当大白鲨啃咬猎物时，它的眼睛会转到脑袋里面。这样可以保护眼睛。

第1名

最快的
灰鲭鲨

灰鲭鲨是最快的鲨鱼。它的游速可以达到每小时20英里（约32.19千米）。灰鲭鲨会跃出水面捕捉猎物。

▶ 第 24—25 页

现在，你看到它们……

一些鲨鱼可以在黑暗中发光！你看到水里有亮闪闪的东西吗？小心！小小的矮灯笼鲨身上有一层可以在黑暗中发光的黏质物。

矮灯笼鲨是一种深海鲨鱼。很多深海动物都会发光。科学家认为，发光可以帮助捕食者吸引猎物。

矮灯笼鲨

▶ 第 26—27 页

现在，你看不到

须鲨

大部分鲨鱼很难被发现。它们的背部是深色的。从上面看，它们与水融为一体。它们的腹部是白色的。从下面看，它们与天空融为一体。

一些鲨鱼用独特的方式来隐身。须鲨的颜色和海床相似。它们的嘴巴的某些部位看起来像海草。鱼游进去后就出不来了！

▶ 第 28—29 页

鲨鱼袭击！

一天，贝瑟尼·汉密尔顿去冲浪。突然，一条虎鲨袭击了她。它用力拖贝瑟尼，她则死死地抓住冲浪板不放。它把她的冲浪板咬下一大块。它还咬下了贝瑟尼的左臂。

在那次袭击之后，贝瑟尼想继续冲浪。她不怕去水里。她知道鲨鱼袭击很罕见。

贝瑟尼说："有一件事从未改变——那就是我冲浪时的心情。"

▶ 第 30—31 页

人类袭击？

鲨鱼袭击非常恐怖，非常可怕。鲨鱼会成为人类的威胁。但人类对鲨鱼的威胁更大。上百万条鲨鱼死在用来捕其他鱼的网里。另一些鲨鱼被蓄意捕杀。

很多种鲨鱼可能将要灭绝。鲨鱼在地球上生活了数十亿年。鲨鱼和人类要学会共享海洋。

灰礁鲨

词汇表

软骨：软骨很轻，很坚韧，富有弹力。鲨鱼的骨骼是由软骨构成的。

灭绝：一种植物或动物不再存在

捕食者：杀死并吃掉其他动物的动物

猎物：被另一只动物杀死并吃掉的动物

史前的：史前的是指在人类把事情记录下来之前的时间段。

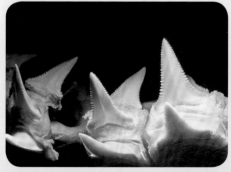

锯齿状的：如果某个东西是锯齿状的，那么它的边缘是凸凹不平的，就像锯条一样。